# Funville Adventures

By A.O. Fradkin and A.B. Bishop

ISBN: 978-1-945899-02-7
Library of Congress Control Number: 2017949993

Text: A.O. Fradkin and A.B. Bishop
Editing: Karla Lant, Maria Droujkova, Dmitri Droujkov
Illustrations and cover: Mark Gonyea
Layout: Howie Severson, Fortuitous
Year: 2017

Published by Delta Stream Media, an imprint of Natural Math
309 Silvercliff Trail, Cary, NC, USA

# CONTENTS

To my students at Main Line Classical Academy and Golden Key Russian School.

—A.O.F.

# 1

# THE THIEF

"Has anyone told you yet that things disappear on that slide?" Lucy whispered to Emmy, pointing to the old, rickety metal contraption at the far corner of the playground. "We call it the Thief." The fourth grade class was just coming out for recess, and the students were very restless from having to sit still all morning. The wind howled through the trees on this crisp, fall day, stirring the fallen leaves. It was the perfect setting for a real-life ghost story, and the old slide often served as inspiration. The new kid in the class, Emmy, had not yet heard the stories of the slide, so the children gathered around to tell her.

"I heard a kid lost his Harry Potter lunch box going down the slide and it was never found," a little boy named Jeremy added, "like the Thief just ate it!"

"And a butterfly hair clip!" Another student chimed it. "Disappeared as if it was yanked straight off the girl's head!"

"Nonsense," Emmy declared. Her younger brother, Leo, whose class had also just joined recess, drifted over to listen as well.

"There's a perfectly logical explanation for everything," Emmy asserted confidently. "Did any of you actually witness this supposed lunchbox disappearance? Why would someone even take a lunchbox down a slide?"

"Maybe he forgot he was still carrying it?" Jeremy suggested.

"But then he remembered that he lost it on the slide?" Emmy countered. "Seems implausible to me. And a hair clip could just fall out on its own and get lost under some leaves."

"Other things have gone missing too," Lucy insisted. "More than can be easily explained. I think it's a magic slide. Maybe an evil magic slide."

Emmy shook her head. "There's no such thing as magic," she said.

Leo looked at her aghast. "How can you say that!" he cried. "Of course there is such a thing as magic!"

Leo was five, and believed fervently in many magic things, like Santa Claus, unicorns, and the Tooth Fairy. Emmy saw his insistent, sincere distress and decided to soften her stance on magic. Even though she felt that at nine, she was older and wiser, and no longer believed in such childish fantasies, she knew Leo took great joy from his belief in magic.

"I take it back," she said thoughtfully. "There may be magic in the world elsewhere. But I'm sure this playground is perfectly ordinary, and there is a non-magical explanation for everything that's gone missing. It hasn't been stolen by some magic evil slide."

"If you're so sure it's not evil, then why don't you go down it?" Lucy challenged.

Silence fell amongst the huddled children. This was a bold dare. No one had actually gone down the slide for as long as any of them could remember. All the rumors of the Thief's mysteries had been passed down as schoolyard gossip for years. It had become more of a myth than an actual slide. Its surface was covered with layers of dirt, cobwebs, and rust.

Looking at it, Emmy hesitated for a moment. The wind howled again as if in warning, but she steeled herself. "Fine," Emmy accepted. "I will go down the slide."

"You have to take something with you," Jeremy said. "To see if it disappears."

The other children nodded. They were anxious to finally see a test of the powers of the mythical slide.

"I will take my notebook," Emmy agreed, extracting it from her school bag and holding it up for all to see. Leo knew this was a brave choice, as Emmy loved to write stories in her notebook, and it was very precious to her.

As Emmy climbed the rusty old slide, notebook firmly in hand, Leo began to get nervous.

"I don't want you to lose your notebook, Emmy," he said worriedly. "I know how much you love your stories, and I love it when you read them to me."

"Don't worry," Emmy reassured him, almost at the top. "I'm not going to lose it. It is going to stay right here in my hand."

Emmy paused at the top and waved the notebook again for dramatic effect. All of the children's eyes were locked on her as she brushed away a few spider webs. Finally, she sat down and pushed off.

"There can be magic, Leo," she said, slowly moving down the rusty slope. "But really, there is no such thing as an evil sl-"

And before she had the chance to finish the word, Emmy disappeared.

Leo looked on in frozen horror at the empty air where Emmy had just been.

# HiDE AND SHRiNK

Leo quickly resolved that, whatever magic this was, he wasn't going to let any thief take his sister away from him. Before anyone could stop him, he ran up the slide and began sliding down after her.

The Thief was not very tall, so even though Leo felt like he was moving down at a snail's pace, he rapidly approached the large pile of leaves covering the bottom of the slide. Just before he expected to come to a complete stop, Leo felt a sudden pull and found himself spiraling down a much smoother slide and at a much faster pace.

This second part of the journey was over even more quickly than the first. Leo did not have time to get scared. He leapt off the edge of the slide and looked around frantically, soon finding Emmy, who was standing nearby, seemingly frozen in bewilderment. Leo was much relieved to see her, and also grateful to have landed on firm ground, which, to his amazement, was now bright orange!

"Are you ok?" Leo asked Emmy, who seemed very surprised to see him.

"I think so," Emmy said, her notebook still clasped firmly in her hand. "Did you follow me down the slide?"

Leo nodded. "Emmy, I think the Thief stole *us.*"

Emmy opened her mouth to say that there had to be a perfectly logical explanation, but then felt that she could not say it with her usual confidence and closed her mouth without speaking.

Leo was feeling apprehensive, but also extremely curious. He started surveying the new surroundings. He noted that they were on a playground not much different from the one they were on a few minutes earlier. It was not much bigger or much smaller and it had many of the same attractions: swings, seesaws, slides. However, there was no doubt that it was not the same playground. For one thing, where there used to be a red pig and brown cow on springs, there were now two gray elephants.

"Where do you think we are?" Leo wondered.

"I'm not sure, but I plan to find out," replied Emmy, regaining some of her self-assurance. "One thing I can say for certain though," Emmy pointed to the slide that they had just come down, "That is not the Thief." There was no doubt about that. This slide was so clean it sparkled in the sun.

As Emmy and Leo continued to walk around the orange grounds, Emmy suddenly stopped and pointed to a strange figure in the distance. "Am I seeing what I think I'm seeing?" she asked Leo. He followed her outstretched arm.

"Looks to me like a purple elephant, standing upside down on its trunk," said Leo matter-of-factly.

"But how can it balance like that?" Emmy wondered.

Leo shrugged. "This place seems full of magic."

"There should be a better explanation," Emmy insisted. "When I said there may be magic elsewhere, I meant far away from us."

"But Emmy," Leo interrupted, "I think we are *elsewhere.* And very far from home."

At this point, they noticed two boys running in their direction. The boys looked to be a few years older than Leo, although not quite as old as Emmy. They wore bright colors and friendly smiles.

"You guys look lost," said one of the boys. "Would you like to join our hide-and-seek game so that one of us can find you?"

The boys seemed friendly enough, but Emmy was still too rattled to react. Leo, however, quickly decided that two boys wanting to play hide-and-seek was a non-threatening and happy occurrence, and blurted out:

"I love hide-and-seek. Who's it?"

"I was it last time," replied the boy, "so this time it's Harvey's turn. I'm Doug, by the way. What are your names?"

"I'm Leo, and this is my sister Emmy."

"Cool! Well, you better find hiding places before I count to twenty. One, two, three . . . " Harvey started counting as he closed his eyes. Doug set off immediately, while Leo looked around for a place to hide. After a moment, Leo noticed some bushes not too far off and sprinted in that direction.

Emmy was much slower at responding, and before she could even decide if she was willing to play, Harvey finished counting and opened his eyes. He gave her a disappointed look and ran towards Leo's bush. Emmy was sure that Harvey knew where Leo was, because Leo had not yet mastered the art of hiding without leaving some part of him visible. Yet Harvey didn't look behind the bush and declare Leo found. Instead, Harvey started waving his hands at the shrub.

After about a minute of such hand motions, Harvey did finally peek behind the bush. He briefly disappeared from sight, and then Emmy heard a loud, "Oops!" Eager to find out what had happened, Emmy ran towards the hiding place. She spotted Harvey as soon as she got there, but there was no sign of Leo. Harvey was kneeling down and looking at something in the grass. Emmy looked down as well.

It took her a moment to realize what she was seeing, and even then she couldn't quite believe her own eyes. She was looking at a creature about as tall as a chipmunk on its hind legs that seemed to be a miniature version of Leo.

"What in the . . . ? How did this . . . ? Please tell me . . . "

# THE UNINTENDED WISH

"Is this . . . ?" Emmy began.

Harvey interrupted her.

"Yes, this is your brother. But don't look so petrified."

"What happened to him?"

"I used my power on him."

"Your power?"

"Yeah," said Harvey as if it was the most natural thing in the world to have a power. "My power is to halve things in size."

"Halve?" inquired Emmy. "But he is no more than a tenth of what he was!"

"That's because I used it multiple times."

"Why did you use it on him at all?" Emmy wanted to know.

Of course what Emmy really wanted to know was how and why did Harvey have a power in the first place? Where were they? Did other inhabitants here have other powers? And much much more. But there would be time to ask those questions later. For now, she needed to snap into action and help Leo.

"You see," Harvey started explaining, "I thought it would be funny if I shrunk the bush, taking away Leo's hiding place. But I must have shrunk him instead of the bush."

"That still doesn't explain why you shrunk him multiple times," Emmy pointed out.

"Oh, that was because I didn't realize what was wrong. I just thought my power wasn't working so I kept re-trying it."

"So what do we do now?" asked Emmy anxiously.

"That's easy," Harvey exclaimed. "We find Doug!"

"You mean you just want to continue with your game as if nothing happened to Leo?" Emmy chided.

"Of course not," replied Harvey sounding somewhat hurt. "We need to find Doug because he's the only one that can help us."

"Let me guess," said Emmy, "his power is to double things in size?"

"Yeah, how did you know?"

"I didn't. It just seemed like the most logical conclusion," Emmy explained. "Do all people here have special powers?" she asked.

"So you're not from Funville?" asked Harvey. "I thought I hadn't seen you before. Yes, every Funvillian has exactly one power. Are you sure you're not a Funvillian? You sure look like one."

"Well I'm not. I'm a Pennsylvanian."

"What's a Pennsylvanian?"

"A person from Pennsylvania," Emmy explained.

Harvey seemed satisfied with that explanation so Emmy suggested they start looking for Doug.

"Doug is pretty good at hiding," Harvey warned Emmy. "It usually takes me at least ten minutes to find him."

"Well hopefully we can do it quicker today since there are two of us," said Emmy as she began surveying her surroundings for good hiding places. Where would she want to hide?

Suddenly she noticed that Leo, who was still in the grass by her feet, was jumping and waving to get her attention. She picked him up and brought him close to her face. His voice was barely audible, but after a while she made out that Leo was squeaking at her the location of Doug's hiding place. She discerned the words "slide," "we," and "from".

She now remembered that out of the corner of her eye she'd seen Doug heading towards *their* slide after

Harvey had closed his eyes. She called to Harvey and they ran to the slide.

When they got there, however, Doug was nowhere to be found. He was not behind the slide or under it.

"Well, it looks like he's not here. But can I go down the slide, just once, before we look elsewhere?" asked Harvey. "I haven't done it today yet, and it's so much fun."

Emmy gave him an "are you serious?" look, but then she felt tempted as well.

"Fine," she said with a smile, "but only if I get to go first."

She climbed the steps and started going down the slide. Suddenly, she hit something. Or was it someone? "What a great hiding place," she thought, "I'll need to remember it the next time I play hide-and-seek." Because the slide was so twisted, Doug was able to curl into it in such a way that he wasn't visible from the ground.

"What a clever hiding place," Emmy complimented Doug, as she was hit by someone from above. Harvey had not noticed the traffic jam when he decided to come down after her. All three had a good laugh, and then they slowly made their way down.

When Harvey explained to Doug what had happened, Doug was quite amused but not at all surprised.

"Harvey and I run into these things all the time," Doug explained. "It's really great having a brother who can correct your mistakes."

"So where is the victim?" he inquired. Emmy realized that she was still holding Leo. "Hopefully I didn't squeeze him too hard," she worried. She placed Leo on the ground and took a step back.

"How many times did you do it?" Doug asked Harvey.

"I think three," replied Harvey, "but I'm not exactly sure."

"It looks like it was more than that," Emmy hypothesized.

"Well, we can take it one at a time," Doug said.

After three iterations of Doug's power, Emmy immediately knew that Leo was not back to his normal size. He looked like their favorite monkey at the zoo. However, after one more iteration, he was returned to his proper size. Nonetheless, Leo looked very ill-at-ease.

"Are you alright?" Emmy asked him.

Leo nodded. He gestured to Emmy to lean in close, so he could whisper in her ear. "Did I hear them right earlier?" he asked nervously. "Are people here fun villains? I'm scared of villains, even `fun' ones, whatever that means."

"Not fun villains," Emmy corrected. "Fun-vill-ians," she pronounced slowly, "meaning, inhabitants of Funville. There's no need to be scared."

"Phew." Leo was visibly relieved. "I guess it's hard to hear correctly when you're the size of a chipmunk."

"We are just heading home to make lunch if you would like to eat with us," Harvey interrupted. "Our house is right over there," he pointed to a large cottage with a neon yellow roof.

"Yay, lunch!" Leo exclaimed. Emmy realized that she was ravenous too and said, "Yes, we will gladly join you."

The house was only a few minutes walk away, and when they got there they found Harvey and Doug's friend Blake welcoming them at the door. He ushered them into the house.

Soon Leo, Emmy, and Blake were seated at a wooden table, as Harvey and Doug set out plates with sandwiches, pickles, and chips.

"My pickle is quite small," Blake complained, and passed his plate to Doug. Doug waved his hand, and the pickle doubled in size.

"Thanks," Blake said, then took a bite. "I really like pickles," he said to Emmy, who was seated next to him.

Now looking at Blake closely, Emmy noticed something a bit out of place. Clipped to the pocket of his shirt was a plastic blue butterfly.

"Where did you get that butterfly hair clip?" she asked, pointing to it on his shirt. "And why do you wear it on your shirt?"

"Is that not how it's supposed to be worn?" Harvey asked. Emmy had failed to notice before in the midst of all the excitement of Leo's shrinking, but she now saw that Harvey and Doug each had the same hair clip attached to their shirts.

"It's a hair clip," Emmy said. "It's designed to be worn in your hair. But I guess there's no reason why you can't wear it on a shirt. I think it's from our world. I was told that somebody lost one."

"Oh really?" Blake was intrigued. "I found it by the slide on the playground. I love that slide - good things often mysteriously appear around it."

"Like us," Leo observed.

"Yes, like you!" Doug agreed.

Emmy wrinkled her brow in confusion. "But there's one thing I don't understand," she mused. "In the story that I heard, only one hair clip was lost, not three."

Blake nodded. "That's true," he explained. "I only found the one I'm wearing. I didn't know what it was for, but I liked it, so I clipped it to my shirt. But then lots of people saw it and wanted one too, so we made copies."

"Oh," said Emmy, wondering how to make a copy of a hair clip.

As she was mulling this over, Leo chewed his sandwich happily, his legs swinging. "This is really good," he announced. "All of that shrinking and growing has made me very hungry!"

"Did Harvey and Doug use their powers on you?" Blake asked Leo.

"Accidentally of course!" Harvey explained. "It always works out fine in the end."

"Yeah, we make a good team," Doug added smiling.

Leo moved on to his chips, noisily crunching and still swinging his legs. Emmy noticed that Blake's legs didn't quite reach the floor either, but he was sitting much more calmly, politely finishing off his pickle and wiping his mouth with his napkin.

“What is your power?” Leo asked Blake.

Blake did not answer right away, but paused to adjust his glasses, which had drifted down his nose as he has been leaning over his plate. “It may be better to show you rather than explain,” Blake replied. “So I will wait for it to be useful.”

This left Emmy feeling a bit ill at ease, but Leo just shrugged and continued crunching his chips.

“What's your power?” Blake asked Leo in return.

“Eye dught hafa powder,” Leo responded, his mouth full of chips.

Emmy, Harvey, and Doug all chuckled. “What was that?” Blake said.

“I don't have a power,” Leo repeated, clearly this time, then immediately resumed chewing.

“We come from somewhere else,” Emmy explained. “So we are still learning how things work around here.”

"Are there pickles where you come from?" Blake asked. "And sandwiches? And hot dogs? And ice cream?"

"Did someone say something about ice cream?" Leo looked up from his plate. Emmy ignored him.

"Yes, we have all those things too," she told Blake. "We just don't have powers."

"Well, we have powers, and dessert!" Harvey declared, and disappeared from the table for a brief moment, then reappeared with a plate of cookies. He set them on the table, and everyone took one.

"I like it here," Leo said, in between bites of his cookie, his legs still swinging happily.

"What's not to like?" Doug added lightheartedly. "Adventures happen every day around here. And you never know when you might get accidentally shrunk!"

"It is all very exciting," Emmy agreed. "In fact, I'm going to write down everything that happens to us here, so I won't forget any part of the adventure. How fortunate that I brought my notebook when I went down that slide."

Emmy looked down at her notebook. The once-radiant-blue hue of the front cover was faded from heavy use, but all of the pages were filled with neat, careful writing. Taking her pen, which she conveniently always kept attached to her notebook, she began to imagine how she would describe this scene: the five of them sitting companionably around the wooden table over plates of sandwiches, the friendly red-headed Funvillian with the mysterious, yet-to-be-revealed power . . . But of course she would have to tell the story from the beginning, starting with their journey from the Thief and their sudden arrival in this exciting new world, then Leo being shrunk to the size of a chipmunk! It would all make an excellent story, she thought to herself.

But with her pen poised over the notebook, Emmy realized she didn't have any empty pages!

Each page of her notebook, from the front to the back, was filled already with her notes, homework, and other stories. "There's no room," she muttered sadly to herself, her pen wavering uncertainly in the air.

Sitting beside her, Blake heard her complaint. Instantly, he waved his hand over her notebook.

"I made room!" he told her proudly.

Emmy opened the notebook to discover the first page was now blank. For a moment, she did not quite understand. Did Blake somehow add extra pages? Then she began flipping through the pages, slowly at first, then with increasing panic. They had *all* gone blank!

# 4

# THE BOTTOMLESS SUITCASE

"Oh no!" Emmy cried.

Blake was looking at her with concern. "But you wanted more space to write things down," he said, "so I erased the pages. My power is to clean things."

Emmy composed herself as she remembered how Harvey's accidental shrinking of Leo was easily corrected. "That's okay," she reasoned. "We can just bring it all back, can't we? There must be someone who *un-erases* things, right?"

Blake was not looking at her, instead he was staring sheepishly at his shoes. "I'm sorry," he said sincerely. "I really was just trying to help."

It began to dawn on Emmy that perhaps some powers cannot be easily undone. "So there's no one who can un-erase things?" she asked again.

Blake, Harvey, and Doug all shook their heads no. "We've never met anybody like that," Doug confirmed.

As this sunk in, Emmy became quite distressed. "But my stories!" she lamented. "I had been working on some of them for over a year!"

Leo was nodding in sympathy. "She does spend a lot of time working on her stories," he told the others. Harvey and Doug were looking back and forth between Emmy and Blake, seeing that both were upset but unsure how to help.

Blake began to tear up. As he wiped his eyes with a dirty sleeve, he smudged his glasses.

"How come I never get it right?" Blake asked in a voice that was so timid that it was barely audible. "I always have such good intentions!" he added with a bit more volume.

"We all make mistakes sometimes," Harvey tried to comfort him, resting a hand on his shoulder. "It must be hard not having a brother who can help you fix them."

"At least when you make mistakes," Doug told Blake, "nobody ends up in danger of being stepped on!"

Meanwhile, Leo had grabbed another cookie, split it in half, and offered half to Emmy. It was a thoughtful gesture, and she started to calm down as she ate it. Seeing Blake's remorse, she began to feel sorry for him. However, she was still upset to have lost so much work.

"I know you were only trying to help," Emmy said finally, "but I wish you had asked me first."

"I truly am sorry," Blake replied, "I will try very hard to do that in the future. Sometimes I wish I didn't have a power."

"It does seem useful when used carefully though," Emmy pointed out. "Like right now you can use it to clean your glasses."

"You're right," Blake said. He blinked firmly, and his glasses were restored to an unsmudged state.

"And did you clean the playground?" Leo asked. "That slide was so shiny!"

Blake nodded. "I keep the slide clean. Seems like the least I can do for the gifts it provides."

Meanwhile, Emmy looked forlornly at the empty notebook she was still clutching. All that work for nothing! But at least she now had room to record their new adventure. As the others started to clear away the plates from the table, she lifted her pen and began to chronicle the events that had transpired since she and Leo had arrived in Funville.

She wrote about their arrival from the slide, her first impressions of this strange new world, Leo being shrunk down and then back up to size, and the sad fate of her stories. This filled a couple of pages. With that done, she got up from the table, ready for the next chapter of their adventure.

Emmy and Leo left Harvey and Doug's house, excited to explore more of Funville.

Soon they happened upon two new Funvillians, a boy and a girl, each carrying a large bag. The two looked strikingly similar, with the same brown curly hair, the same brown eyes, and even an identical pattern of freckles. Emmy noticed that they were also both wearing the already familiar butterfly hair clips attached to their shirts.

They were carrying heavy luggage, so Emmy and Leo offered to help. The Funvillians happily accepted.

"I am Cory and this is my sister Marge," the boy said.

"We haven't seen you around before," Marge said. "Who are you?"

Emmy and Leo introduced themselves.

"What's in your bags?" Leo asked. "Looks like a lot of stuff!"

The two bags were quite stuffed, their surfaces bulging from the odd shapes of their contents. Leo could see what looked like the outline of a doll on one side, positioned with her arms raised as if she was trapped, pressed against the cloth from the inside and banging her fists to get out.

"It is a lot, but you only know the half of it!" Cory claimed, a mischievous spark in his eye. "You see, we often go camping, which is pretty fun-"

"*Usually* pretty fun," interjected Marge.

"-but sometimes it rains," Cory continued, "and then we have to stay inside a cabin, bored with nothing to do until the rain passes."

Marge sighed and twiddled her thumbs to illustrate the level of boredom. "But this time we've prepared for the rain by bringing our rain gear," she added. "However, we're still bringing *all* of our toys as usual, just in case," she finished with a smile.

Looking more closely at the bags, Leo could now make out the likely outlines of a pack of cards, a baseball, and several rectangular objects that could be books or board games.

"Can I see?" Leo asked.

"We'll show you," Cory promised. He set his bag on the ground and unzipped it. He pulled out a red fire engine. "This is one of our favorite toys," he said, handing it to Leo.

"Cool," Leo said. "But do you have to share it? Emmy never wants to share her toys with me."

Emmy rolled her eyes. "That's because you always break them!"

"I sometimes break them *accidentally*," Leo stressed.

"We don't have to share it," Cory said. "Watch this."

Cory took the fire engine back from Leo, carefully placed it on the ground, and then took a step back. Just to be cautious, Leo took a step back from the toy as well.

All eyes were locked on the fire engine, as Cory pointed at it firmly, a look of deep concentration on his face. Leo blinked, as he couldn't believe his eyes. Suddenly, where there had been just one fire engine, there were two!

"Pretty cool, huh?" Cory said.

Leo was mesmerized. He looked carefully at the two fire engines. They appeared to be identical. There was even a small scratch in the red paint at the same spot on each of them.

"Can I touch them?" Leo asked.

"Sure," Cory said. Leo picked one up in each hand, holding them up for Emmy to inspect more closely.

"Aha!" she said, "So you make copies of things, and that explains the multitude of butterfly clips."

"Yes, that was all my doing," replied Cory, admiringly looking down at the clip on his shirt.

"But wait!" exclaimed Leo, "Now those fire engines won't both fit back in your bag!"

Indeed, he was right. Cory's bag was close to bursting, and even just one more fire engine clearly would not fit. Marge's bag was stuffed just as tightly, with no extra room for the new fire engine either.

"But that's where I come in," Marge said proudly. "Watch what I can do!"

Leo passed the two identical fire engines to Marge. She took one in each hand, holding them up so Leo and Emmy could see. She slowly brought her hand together, and the two fire engines touched and then merged into a single one. She gave the single fire engine back to Cory, who returned it to his bag.

"You see?" Marge said. "We can have two of everything, but we only have to carry one!"

Emmy was puzzled. "But if you can take two toys and turn them into one," she asked, "why can't you keep doing that until your bags each contain only one toy? Why do you still have so much to carry?"

"I can't take just *any* two toys," Marge explained. "They have to be the same. So if I try to do it on a fire engine and a puzzle or something, it won't work. That's why we still have so much in our bags. We need to carry one of everything."

"Can you do it to things that aren't toys?" Leo inquired.

"Indeed," Cory replied, "Remember the rain gear Marge mentioned? Well, I couldn't fit two rain boots into my bag. So I packed only one! I can copy it on the spot if I need to."

"That's very cool," said Leo with a sigh. "I wish I could replace shoes with toys when we go on our camping trips."

Emmy, on the other hand, had a skeptical look on her face. "Have you actually tried copying the boot and trying both on?" she asked.

"Why would I do that?" asked Cory, honestly puzzled by the suggestion. "I've tested out my power on enough things" he added confidently.

"And you've never been surprised by the outcome?" Emmy asked suggestively.

"Not recently," Cory replied proudly. "But it seems to me that another demonstration is in order," he continued as he pulled a bright yellow boot out of his bag.

He placed the boot carefully on the ground and looked around at his audience before proceeding to copy it. Emmy noticed that Marge was staring intently at the two boots. Suddenly, Marge put her hand to her mouth as if to suppress a laugh, then exchanged a knowing look with Emmy.

Meanwhile Cory, oblivious to this silent dialog, started putting on the boots. His left foot slid into one boot perfectly, but the right foot didn't seem to quite fit. Still, Cory forced it on and took a step.

"Why is this so uncomfortable?" he complained, as he looked down at his feet. "Oh!" he chuckled. "Now I see why this may not have been such a great idea."

Emmy smiled.

"But does that mean that you now have to leave some of your toys behind to fit the second boot?" asked Leo with great concern.

"No, no," Marge reassured him. "I'm sure we can think of something."

"I guess I don't need five different shirts," Cory realized. "I can bring one and copy it."

Leo appeared much relieved.

Meanwhile, Emmy imagined other things that could be copied and merged. Suddenly, she got a disturbing thought. "Cory," she asked, "can you copy people? Like, could you make two of me? Or Leo? Or Marge?"

Marge could tell that Emmy was troubled, and reassured her. "Don't worry, Emmy," she said. "Our powers only work on inanimate objects."

"What does `inanimate' mean?" Leo asked.

"Non-living," Marge clarified. "Cory and I can't apply our powers to living things."

Emmy let out a sigh of relief. The thought of her insides merging together with those of another copy was not a pleasant one. Leo, however, was slightly disappointed.

"It would have been cool to have two of me," he lamented. "Then one could be doing chores while the other played!"

Cory chuckled. "Sorry, friend," he said. "No such luck. But don't worry, you won't have to do chores here in Funville. Life here is more like one big party."

"Oh," interrupted Marge, "and there is a party tonight. It's a birthday party for our friend Ida. You should both come, it's going to be very fun."

"In fact, we should head over and help prepare." Cory added. "Would you like to join us? It's at the school, just a few blocks down the road. You will have a chance to meet many new Funvillians."

"Sounds great," agreed Leo and Emmy in unison.

At this point they reached Cory and Marge's car and loaded the suitcases. "Where do you go camping?" Emmy asked. "Is it outside Funville?"

"No, it's still in Funville," Marge responded. "Just far enough away that we need a car to get there."

Emmy couldn't help but show a bit of disappointment.

"Why?" Marge asked. "Do you want to leave Funville?"

"We are having a great time in Funville," Emmy reassured her. "But at some point we should return home, and I don't know how to get there."

Marge furrowed her brow. "We've never left Funville," she said. "And I don't know of any Funvillian who has."

Suddenly Emmy got a sinking feeling. It dawned on her that if some powers couldn't be reversed, perhaps their trip to Funville couldn't be reversed either, and they might be stuck forever!

# UNEXPECTED PRESENTS

As Emmy and Leo followed Cory and Marge into the school, the party preparations were already well underway. There were many Funvillians scattered around the large gymnasium, chatting in groups. All around the room were boxes and wrapping paper for the presents, as well as colorful signs and banners to decorate the walls. The Funvillians started noticing the new arrivals and several of them chimed in unison, "Yay, Cory is here! Now we can make many more balloons!"

Leo was confused. "Cory, you can make balloons?"

"They already have a few balloons," Marge explained. She gestured to a small bag of uninflated balloons of various colors. "So Cory can replicate them many times, and then we will have lots of them!"

"I'll get started," said Cory, as he took the bag of balloons and pulled out one of each color.

"There are a lot of people here," Emmy said.

"Well, there are twenty-seven Funvillians in here, counting you two, whom I don't have the pleasure of knowing," replied a voice from behind Emmy and Leo. Emmy and Leo turned around to see a new Funvillian who had come in right behind them.

"How were you able to count everyone so quickly?" Emmy asked bewildered.

"That's because she's Connie," supplied Cory, "and her power is to count things."

"How is that a power?" asked Leo. "I can count things too." And Leo proceeded to show off his counting skills. He didn't get very far, however, before he was interrupted by Marge.

"You see," she explained, "Connie can count things *very quickly*. She can look at a group of objects and immediately tell you how many there are."

"As long as they are all in my field of vision," Connie explained.

"It seems like a very useful power to me," said Emmy. "It is great to meet you, Connie. I am Emmy and this is my brother Leo. We are not from Funville so we are still learning about how things work around here and what types of powers different Funvillians have."

Meanwhile, Cory had made several balloons and was passing them to Marge, who was struggling to blow them up as quickly as Cory was making them. Observing this, Emmy had an idea.

"Hey Cory," she called to get his attention. Cory looked up from his work.

"Why don't you blow up the balloons *before* you copy them? Then Marge won't have to inflate them all one by one. You'll just have to inflate one of each color!"

"That's an excellent plan!" Marge exclaimed gratefully. "That will save me a lot of work!"

"Why didn't I think of that?" Cory added, impressed by Emmy's insight. He blew up a single balloon of each color and then began to copy them. As they floated up towards the ceiling, the room quickly filled with color, creating a truly festive atmosphere.

"You two should mingle and meet all of the other Funvillians," Cory suggested to Emmy and Leo. "Visitors from outside of Funville are rare and everyone will be excited to meet you."

"We are excited to meet them all too!" Leo replied, grabbing Emmy's hand and eagerly pulling her into the crowd.

At a glance, Emmy thought the first two Funvillians they approached were siblings. In some ways, they couldn't have been more different. One was a tall thin blonde while the other one had dark hair and was much shorter and rounder. But it was their eyes and facial expressions that gave them away as being close relatives.

They introduced themselves as Heather and Liza and proudly showed Emmy and Leo their present. It was a ballerina doll that balanced on just the toes of one foot.

Emmy was astonished. "How does she not fall down?" she asked, staring at the doll in amazement.

"But that's so easy," replied Heather. "You see, we made her leg very heavy and the rest of her body very light."

"Oh! Did you use that same trick on a purple elephant near the playground?" Emmy inquired.

"Yes, we did," Liza said proudly. "That's how it stays on its trunk."

"I see," Leo commented. "I thought it was pure magic."

"It's magical indeed," Emmy quipped, "But now the magic makes sense!"

"So your powers are to make things heavier and lighter," she continued. "I can see how you could get very creative with that."

"Exactly," said Liza and Heather in unison and then looked at each other and giggled. "They must not know about jinx," Leo whispered to Emmy.

Next Emmy and Leo encountered Harvey, Doug, and Blake, who greeted them as old friends might do, with a hand wave and a very enthusiastic hello. Harvey and Doug's present was an oversized gingerbread house. They were struggling to fit it into a box.

"I told you that you made it too big!" accused Harvey.

"Oh fine," Doug pouted. "Go ahead, make it smaller. But it won't be nearly as much fun then."

"You can always make it big again after she opens it!" Leo reminded him.

"Good point!" Doug said.

Blake was also eager to show Emmy and Leo his present.

"I found this by the slide," he explained, producing a Harry Potter lunchbox. "I think Ida will like it. Do you know the boy on the box?" he asked Leo, tapping Harry Potter's picture. "Is he a friend of yours?"

"Oh I wish!" Leo responded. "He isn't real. He's a character in a story. In the story he has special powers, kind of like a Funvillian!"

"Cool!" Blake replied. "I think Ida will like it then."

But Leo was puzzled by something. "Harry is missing his scar," he remarked.

"His what?"

"His scar," Leo insisted, pointing to the place on Harry's forehead where the lightning scar should be.

"Oh, that!" Blake remembered. "It looked funny, so I cleaned it. Looks much better now, right?"

"Sure," Emmy said, not wanting to contradict him. "Looks great."

Emmy and Leo then wandered over to a brother and sister, Stanley and Carly. Stanley was packing their present into a box, along with a hand-written card. He tied the box shut with a colorful ribbon. Carly waved her hand over the knotted ribbon, and its loose ends curled into a festive decoration.

"My name is Carly and I can curl things," she explained to Emmy and Leo, pleased with her work. But then she suddenly remembered something.

"Oops," she said to Stanley, "I forgot to write something important on the card!"

Stanley sighed and waved at the ribbon. The ribbon wiggled itself untied and fell to the ground in a flop, like a wet noodle. "My name is Stanley and I straighten things," he told Emmy and Leo. "Which means I can quickly untie knots." Carly retrieved the card, wrote her previously forgotten sentiment, and returned it to the box. They tied and curled the ribbon all over again.

Leo then noticed a black-haired Funvillian alone in a corner of the room, clutching a toy stuffed elephant. She was wearing a blue dress with little grey elephants on it. "What is your name?" Leo asked as he approached her.

"Constance," she said, as she placed the stuffed elephant in a box.

"And what is your power?" Leo inquired.

"Elephants," Constance said simply. Emmy and Leo waited for more of an explanation, but none seemed to be forthcoming.

"Elephants?" Emmy asked. "We certainly *have* seen a lot of those. But what exactly is the power?"

"I turn everything into grey elephants," Constance explained. To demonstrate, she pointed to the bows decorating her shoes. Suddenly they were little elephants instead, like the ones adorning her dress. "See?"

"I see," Emmy said. She remembered that the upside-down elephant by the playground was purple. "But why was that playground elephant purple?" she asked.

Constance shrugged.

"I like elephants," Leo added.

Constance smiled at him. "So do I," she said.

"Does Ida like elephants, too?" Emmy asked.

Constance shrugged.

The boxes were now all stacked up against the wall. Everybody in the room turned to watch an ordinary-looking Funvillian that Emmy and Leo somehow hadn't noticed before. She slowly approached the boxes.

"Who's that?" Emmy asked Constance.

"Ivy," Constance told her.

"What's her power?"

"You'll see," Constance said.

"It looks like she is going to use her power on the boxes," Leo whispered to Emmy.

After a bit of suspense, Ivy waved her hand at the boxes and . . . they all turned green.

"Nice! They look much more festive now!" Leo exclaimed.

But Ivy did not seem to share his sentiment. In fact, she looked rather shocked.

# THE BENEVOLENT PRANKSTER

Ivy waved her hand again, and this time the boxes flipped upside down.

“Better be careful,” warned Emmy. “Some of those things looked pretty fragile.”

Leo meanwhile was ecstatic. “I didn’t know that a Funvillian can have multiple powers!” he marveled.

Ivy seemed to realize something. Her expression shifted from confusion to annoyance. She began scanning the room as if looking for someone. Seemingly not finding the right Funvillian, she turned to the boxes once more. This time, a wave of her hand caused the boxes to shrink in size, and everyone giggled.

"Hey, that's Harvey's power!" exclaimed Leo, but before he could even finish the phrase he was interrupted by an almost furious Ivy.

"Fay," she bellowed, "where are you? I know you're hiding in here somewhere!"

From behind a curtain appeared a short Funvillian with bright green hair and very mischievous eyes.

"Sorry, Ivy!" Fay said lightheartedly. "I just couldn't resist the temptation to show off my power to the new visitors."

Leo was very confused. "So she doesn't have *all* of those powers?" he asked, pointing at Ivy.

But Emmy began to understand. "Was that you changing her powers?" she asked Fay.

Fay winked at Leo, who still looked absolutely baffled.

"Yes," Fay said. "I can change a Funvillian's power just for one use. So when I stop, the original power returns."

After pausing to think this through, Leo clapped his hands in excitement. "Does that mean you could temporarily give me a power?" he asked.

"Let's try and see," Fay said adventurously.

"I want Cory's power so that I can make my army of toy soldiers as big as I want!"

"Ok, let's try," she said, and dramatically waved her hands at him. "Now go try it out on that chair over there," Fay suggested.

Leo walked grandly up to the chair and started waving his arms theatrically at it.

As if unimpressed, the chair did not react. The Funvillians burst out laughing, Fay loudest of all.

"Am I not doing it right or something?" Leo asked uncertainly.

Emmy looked at Fay suspiciously. "Did you know that wouldn't work?"

Fay winked. "I guess my power only works on Funvillians."

Leo was disappointed, but said good-naturedly, "Well, it was worth a try!"

Ivy tapped Fay on the shoulder. "Now that you've had your fun, will you stop interfering so that I can finish what I set out to do?"

"Okay," Fay said. With Fay's help, the boxes were returned to their original condition, except everyone agreed to leave them green, as it was indeed more festive! Ivy returned to the wall of boxes. With one eye locked on Fay, she waved her hands at them yet again.

This time, the boxes disappeared.

"Where did they go?" Leo asked.

"Don't worry, they are still there," Ivy assured him. "You just can't see them anymore."

Meanwhile, Emmy had been deep in thought. She turned to Fay and asked, "Can you change a Funvillian's power to *anything*?"

"Not quite," she paused mysteriously. "Why do you ask?"

Emmy told Fay the unfortunate fate of her notebook. She asked if there was a power that could bring her stories back.

Fay shook her head. "Un-erasing is not a power a Funvillian can have. That would be too much like magic."

"I don't understand," Leo complained. "Harvey and Doug can undo each other's powers, as can Cory and Marge. So why can't Blake's power be undone?"

"Blake's power is different," Fay explained. "When Cory applies his power to a red truck, he gets two red trucks. When he applies it to a brown teddy bear, he gets two brown teddy bears. So when Marge sees two brown teddy bears that Cory produced, she knows that they could only have come from one brown teddy bear. Now when someone sees a blank page that Blake erased, they have no idea what was written there before he did that. It could have said 'My name is Emmy,' or 'My brother is awesome,' or 'homework is so annoying.'"

"It could even have had a picture or just scribbles," Leo supposed.

"Exactly," said Fay.

"I understand," said Emmy sadly. After thinking for a moment, she continued, "I bet Constance's power can't be reversed either, since you can't know what the object was before she turned it into an elephant."

"Ooh, you're quick," Fay clapped her hands.

"Well, I'm starting to learn the rules of how things work here in Funville," Emmy said. "I think at this point I know what to expect."

At that moment, a purple dog appeared in the doorway.

# A COLORFUL SURPRISE

Just behind the dog, a new Funvillian entered. "Am I late?" he asked.

"You're too late to help with the preparations, and too early for the party!" Fay answered.

"This is Randy," she told Emmy and Leo. "His power is particularly interesting, as it changes from day to day! Since there is still plenty of time before the party, Randy, why don't you show Leo and Emmy more of Funville and tell them about your power?"

"Sure!" Randy agreed. "Come with me!"

Emmy was still eyeing the purple dog, wondering why no one else had given it a second glance.

"Is this your dog?" she asked Randy.

"Oh, that's Tess," Randy replied. "She'll come with us."

Randy led Emmy and Leo out of the gymnasium. "My power has to do with colors," he began to explain. "But I can't fully control it. Each day I turn things a different color. One day, I tried my power first on my shirt, hoping for a cool color, but it ended up puke green! Some of the Funvillians laughed when they saw me, as I looked a little bit silly. It was so embarrassing, I just wanted to hide out for the rest of the day until I could change it the next day. But Ida saw me next, and she took a green marker and drew on her shirt to make it look puke green too. 'I think it's very fashionable!' she said, and no one else laughed at us for the rest of the day. She is a great friend."

"She sounds very nice, I am looking forward to meeting her at the party," Emmy said. The dog bounded along beside them, her tail wagging merrily. Randy pulled a small dog treat out of his pocket and fed it to her, and she chewed it happily.

"Hey, Tess is the exact same color as that purple elephant we saw!" Leo exclaimed excitedly.

"Oh yeah," Randy recalled. "I did stop by that elephant today."

Leo reached down to pet the dog, who leaned fondly into his hand.

"Tess is very friendly," Randy continued, scratching behind her ears. "But I am her favorite Funvillian. We have been almost inseparable since we first found each other. She isn't always purple of course," Randy continued. "I often use her to test my power for the day before applying it to anything else. So some days she is neon green, or ocean blue, or burnt orange. Well, you get the idea."

"She's very cute," Leo said, as Tess rolled in the grass and then turned her large, sweet puppy eyes up towards him.

"That's one of the great things about her," Randy said. "Everything looks good on her! Colors that would look silly on my shirt or in someone's hair or on a wall, still all look adorable on Tess."

"What did you do before you met Tess?" Emmy asked. "How did you test your power each day?"

Randy laughed. "I would just test it on arbitrary things at first, but that turned out to be a not-so-great idea," he admitted. "This one time," he continued, "I tested my power on a traffic light! Turned it red for the whole day!"

Emmy understood why this would be problematic, but Leo didn't. "What's so wrong with that?" he asked.

"Well, red means `stop,'" Emmy explained. "So if it just stays red, people will never think it's safe to go and . . . "

"They will just be stuck there!" Randy finished, chuckling at the memory. "It was a mess. All the cars stood still, just waiting for the light to change. They waited and waited and waited, until finally somebody got out and started directing traffic. I wanted to fix it," he said earnestly. "But what could I do? The only thing I could do that day was turn things red!"

Tess fell into step beside Leo as they continued walking. "What color is she naturally?" Leo asked Randy.

"I actually don't know," Randy explained. "On the day that I met Tess, I was out walking in the woods. And I was very confused, because my power didn't seem to be working at all. I kept trying it on the grass, on the leaves of the trees, and even on my shirt. And nothing happened! So I tried it on a bush, and still nothing! But then Tess comes marching out of the bush, and at first I think, 'Wow, I've never seen a dark green dog before!', but then I realize - my power that day was to turn things dark green, and I'd only been trying it on things that were already dark green! That's why it seemed not to be working."

Leo thought this through. "So you happened to be wearing a dark green shirt too?"

"Yes," Randy said. "The same exact shade of green as the grass and the leaves. It was quite a coincidence. So I never saw what color Tess is naturally, unless of course, I'm wrong and she actually is green!"

Emmy chuckled at the thought.

"Oops!" Randy suddenly exclaimed. "I still need to finish my gift for the party! I need to go home and get it."

"You came late *and* you forgot your gift?" Leo teased. "Do you always forget things?"

Randy blushed.

"Leo," Emmy scolded. "That isn't very polite."

"Sorry!" Leo said. "I didn't mean to hurt your feelings. I forget things all the time myself!"

"No problem," Randy reassured him. "No offense taken. I am easily distracted. But it isn't just that I forgot. I haven't been able to finish it for a reason. You see, it's a picture of sunflowers, which are Ida's favorite, but I need to make the petals yellow, and my color of the day hasn't been yellow since I started working on it. I've got the green for the grass and the stems and the blue for the sky and the white for the clouds, but I'm still missing the yellow!"

"Hmm . . . ," Leo thought for a moment. "But don't you have a yellow marker or crayon or something at home?"

Randy laughed, amused at himself. "You're absolutely right, I do have yellow markers and crayons at home! I got so focused on using my power for the coloring, I forgot there are other ways! Thanks for reminding me."

"You're welcome," Leo said proudly. "Glad I could help!"

"You can head home to get your gift," Emmy said. "We can easily get around on our own," she assured him. "And don't get sidetracked along the way!" she joked.

"I will try my best!" Randy replied with a wink, as he departed. Tess bounded off after him.

Emmy and Leo continued walking, chatting excitedly about their adventures so far. "Look!" Leo exclaimed, stopping suddenly in the middle of the path. "Have you ever seen a tree upside down like that?"

Emmy's eyes followed the direction of Leo's outstretched arm, and she saw a large tree with its branches touching the ground and its roots reaching up to the sky.

# A LEVELED PLAYING FIELD

As they walked towards the tree to get a closer look, they saw two girls dressed in baseball uniforms. One was reaching into the branches of the upside down tree to retrieve a baseball.

"Oh, hi there!" the other one said, as she noticed Emmy and Leo approach.

Too mesmerized to respond, Leo continued staring at the tree. Emmy had recovered enough composure to ask, "How did this tree get upside down?"

"Oh that?" the girl who greeted them said. "Well, I did that, of course. I'm Rosalinda, and this is my cousin, Roberta." She indicated her companion, who was emerging from the branches with the ball in her hand.

"But wait!" said Leo, waking from his stupor. "I still don't understand how this tree got upside down."

Rosalinda chuckled. "I'll show you, just take a step back."

Emmy and Leo obediently backed away to a safe distance. Rosalinda waved her hand in the air, and the tree began to spin. It landed right side up again, with its roots in the ground and its branches reaching towards the sky.

"That's a cool power!" Leo exclaimed.

"It comes in handy," Rosalinda explained. "Roberta hit the ball into the tree, and it got stuck in the branches. So I rotated the tree upside down so we could get it back."

"What about you?" Emmy asked Roberta. "What is your power?"

"I rotate things too," she said. "But a little bit differently from Rosalinda. Here, I'll show you."

She motioned towards two bicycles that were leaning against a different tree. Emmy and Leo watched as one of the bicycles rotated vertically into the air, with only one wheel still touching the ground.

"You see?" Roberta continued. "Rosalinda rotates things halfway around. So if she does it twice, they end up back how they started. I rotate things one quarter of the way around, so I have to do it four times before it gets back to how it was. We also have two other cousins, Robin and Roy. They also rotate things, but need to use their powers even more times to go all the way around. It's easy to keep us all straight, because Rosalinda rotates things the farthest, and she has the longest name. And I rotate things the second farthest, and I have the second longest name, and then Robin, and then Roy, who rotates things by the smallest amount.

"I see," Emmy said. She pondered their respective powers for a moment. Turning to Roberta, she observed, "So since you can rotate things by one quarter and Rosalinda rotates them by one half, you can do everything Rosalinda can do and more."

"That's right," Rosalinda admitted, "but I can do my part faster!"

Meanwhile, Leo was still gazing at the vertical bicycle with its front wheel staying implausibly aloft, and noticed something unusual. "That bicycle has no pedals," he pointed out.

Emmy turned to look at it more closely. "You're right!" she agreed.

Leo asked, "How can you ride those without pedals?"

Rosalinda and Roberta exchanged knowing smiles. "Well," Rosalinda began, "when you ride a bicycle, what do the pedals do?"

"They go round and round as you move your feet," Leo quickly replied.

"Yes," Rosalinda agreed, "but why do they need to go round and round?"

Leo thought about this for a moment. He imagined his own small bicycle at home, the one with a cool red stripe and black handlebars. He imagined pedaling it down their street, with the wind on his face.

"The pedals make the wheels turn," he said. "So a bike has to have pedals!"

"What about when you are going down a steep hill?" Roberta chimed in. "You don't need to turn the pedals then, right? Because gravity gets the wheels turning even without you turning the pedals."

"Yeah, that's true," Leo said thoughtfully.

"So really what you need," Roberta summarized, "is *something* to make the wheels turn. We don't need pedals for *that* even without a hill, because we can rotate the wheels with our powers!"

"Wow, that is so awesome!" Leo exclaimed.

"We like to race our bicycles, and I always win," Rosalinda bragged.

"Aha! This is where rotating them faster must come in handy!" Emmy realized.

"And poor Roy is always last," Roberta explained sympathetically. "Even when we are not racing, he is always calling, 'Wait for me!'"

"Hmmm," Emmy mused. "I wonder if something can be done to give him a fairer chance."

"I wouldn't mind more competitive races," Rosalinda conceded, "it would be more fun. But I'm just faster! And it wouldn't be fun if I just let the others win."

"That's not what I mean," Emmy continued. "I mean . . . " she paused to think. Rosalinda, Roberta, and Leo looked at her expectantly. "Well, I'm not sure what I mean yet," she said. "But it'll come to me."

"Would you like to meet our other cousins?" Rosalinda asked. "We are about to head home and see them."

"Sure," Emmy replied. "I would like to meet them."

Since Emmy and Leo did not have bicycles, Rosalinda and Roberta politely walked their bicycles along the path to their house as they all continued chatting. When they arrived at the house the cousins shared, Leo and Emmy at first perceived it to be quite normal, very much like their own home. There were some toys strewn along the floor, some paintings decorating the walls, and a large table in the center of the dining room.

But something was a bit strange, Emmy slowly realized. While the pictures and paintings on her own walls at home featured a variety of shapes and subjects (like trees, flowers, and her family members), all of the paintings here were of circles, and were perfectly symmetric!

"Interesting decorating," she remarked.

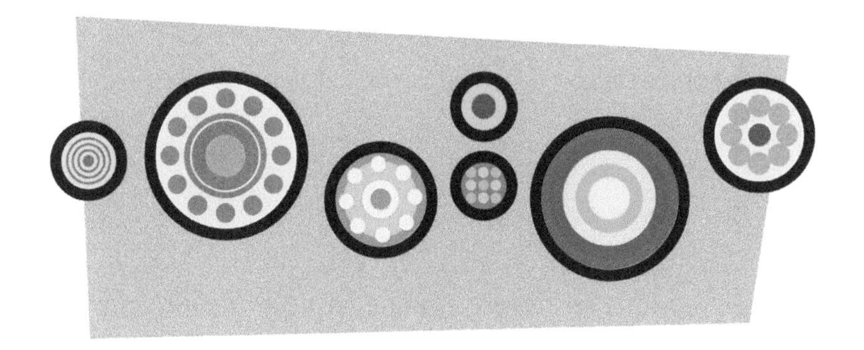

“Yeah,” Roberta said. “We used to have other paintings hanging on the walls, but we kept annoying each other by turning them to different angles.”

“I really liked the dog picture upside-down,” Rosalinda chimed in nostalgically. “But Roy liked it better right side up.”

“And Robin liked it halfway between, so it looked like it was standing on its tail!” Roberta continued. “So really there was no rest. The poor pictures were constantly being turned around and around. We decided to stop the madness and got these paintings that look the same no matter how you turn them.”

“I sometimes miss the dog standing on its tail,” a new voice said, coming from a nearby staircase.

Just then, two more Funvillians came into view at the top of the stairs.

"I'm Robin," the first one said, bounding down the stairs to meet Emmy and Leo.

The second one slid down the railing, arriving just a moment before his brother.

"And I'm Roy," he declared. "My cousins must have already told you how I'm the slowest, but what I lack in speed, I make up for with zeal and charm." He smiled warmly at them.

"Well, I'm glad you don't lack for self-confidence," Emmy good-naturedly teased.

Reminded of the bicycle races, Emmy thought again about the problem of making them more fair. "I've got it. The wheels!" she exclaimed.

"Huh?" Leo tilted his head in confusion.

"I've got an idea for how you can win some of the bicycle races!" she told Roy.

"Really? How?" he asked excitedly.

# THE ELEPHANT IN THE ROOM

"Think about the wheels," Emmy began to explain. "How fast you go is affected by how fast you rotate the wheels, but it is also affected by the *size* of the wheels! If Rosalinda uses a bike with smaller wheels, she will have to turn them faster to compensate, and then the race can be more fair. Rosalinda could use the smallest wheels, Roberta's wheels could be larger, and then Robin's wheels could be even larger, and your wheels could be the largest of all, Roy! That way it could be even, since Rosalinda now has to turn the small wheel more times to cover the same distance."

"That's an excellent idea!" Roberta chimed in. "We can get Harvey and Doug to adjust the sizes of our wheels. That will make our races so much more suspenseful!"

"But for now," Rosalinda interjected, "it's time for us to head to the party. There will be lots of fun games there."

"Will I be able to play too?" Leo asked, as they all left the house together.

"Of course," Robin assured him. "I bet you'll really enjoy them!"

Together they made their way to the party. Entering the gymnasium for the second time that day, Emmy was impressed by the festive decorations and the cheerful atmosphere. The sounds of laughter and playful scurrying of the many Funvillians in attendance filled the large space, making it feel welcoming. Emmy and Leo were both very curious to meet Ida, the reason for all of this fanfare.

In the midst of all the bustling energy, there was an unassuming Funvillian with red hair and a shy smile. She waved at Emmy and Leo, who crossed the room to greet her.

“Hi,” she said, “You must be Emmy and Leo. I’ve heard a lot about you two. I’m Ida.”

“We’ve heard a lot about you too,” Emmy told her. “We’re very excited to meet you.”

“And I’m excited to meet you too! I’ve been told that you are both very wise,” Ida continued, “that you have helped several Funvillians sort out their troubles. Marge told me how you, Emmy, saved her from having to blow up a hundred balloons, and Randy told me his gift is only finished because of you, Leo.”

Leo blushed.

"And I can add that Emmy discovered a way for me to have a real chance to win a bicycle race!" Roy chimed in excitedly, as he snuck up behind Emmy and Leo.

"We have been able to figure some things out," Emmy conceded. "But there is something important I've been trying to figure out and failing. As much fun as we are having in Funville, we will eventually need to go home. And I don't know how to get back there!"

Ida was sympathetic to Emmy's plight, but did not share her concern. "I wouldn't worry if I were as clever as you," she reassured Emmy. "I'm sure you will figure it out."

Emmy nodded to accept Ida's vote of confidence, but remained unconvinced. As she tried to recall everything the other Funvillians had told her about Ida, she realized that no one had mentioned what Ida's power was. She began trying to formulate a polite way of inquiring, but Leo got ahead of her and just blurted out, "So what is your power?"

Ida smiled. "I admire your irrepressible curiosity," she told Leo. "We don't get too many visitors here, so I often forget that not everyone already knows. My power is to keep things the same."

"Oh," Leo said, betraying a hint of disappointment, but Ida didn't seem to notice. She was instead gesturing to the activities swirling around them. "I hope you enjoy the party!" she said. "We have plenty of food and games and it should be a lot of fun!"

"Thanks," Emmy said, as Ida turned to greet some newly arriving guests.

"To keep things the same?" Leo whispered to Emmy. "Doesn't sound like much of a power to me."

"Shhh," Emmy said. "Don't be rude."

Surveying the room, Leo then noticed a large display of ice cream, with several Funvillians helping themselves to generous portions. "Is the ice cream for everyone?" he asked, blushing a bit.

"Of course!" Ida replied. "Help yourself! We have chocolate, strawberry, vanilla, mint chocolate chip, cookies and cream . . . "

Before Ida could even finish listing the flavors, Leo was racing off towards the ice cream.

"Cookies and cream is his favorite," Emmy explained. Ida and Emmy watched as Leo carefully plopped three scoops of cookies and cream into a dish.

"He must really love ice cream!" Ida said.

"Indeed," Emmy responded. "There's no limit to how much of it he can eat!"

As Leo was finishing off his first scoop, a group of Funvillians across the room called out for the others to come and play a game of musical chairs. Not willing to miss the fun, Leo set his ice cream down on a nearby table and ran off to join the game. Emmy and Ida followed more leisurely in his wake.

Cory, Marge, Constance, Ivy, Vicky, Harvey, Rosalinda, Ida, and Leo all agreed to play.

"We need someone not playing to referee," Cory declared. "Who is willing to take the job?"

"I'll do it," Emmy volunteered. "Why do you even need a referee for musical chairs? Isn't it pretty clear how to follow the rules?"

Ida smiled mysteriously. "You'll see," she told Emmy. "It's actually a tough job!"

"I'll help you," Connie offered. They began by arranging eight chairs for the first round, as there were nine players.

The music started. The players began circling the chairs in an orderly fashion, listening carefully for the music to stop. When it did, there was a mad scramble towards the chairs. With the large number of anxious and quick-footed players, it was hard for Emmy to follow the action, but the dust quickly settled and all players were seated. There was a slight pause, as Emmy realized what was wrong.

"Wait," she said. "How is it possible that *everyone* found a chair?" She looked to Connie for help.

"There are now nine chairs," Connie said, "but we started the round with only eight."

Emmy scanned the faces of the Funvillians, looking for a guilty party. Constance, who was seated on a chair next to Cory, raised an eyebrow suggestively in his direction. Cory was looking down at the floor, not meeting Emmy's gaze.

"Cory?" Emmy asked, in a tone that was her best imitation of a teacher who has caught a student cheating on a test. "Did you copy a chair?" She placed her hands on her hips and stared at him accusingly.

Cory was unable to keep a straight face and broke out laughing. "Fine, you caught me!" he admitted. "I saw I wasn't going to make it to the last chair before Constance, so I copied her chair."

"I assume he should be disqualified?" Emmy looked to Connie for confirmation. Connie nodded.

"Disqualified!" Emmy declared, pointing at Cory. Shuffling his copied chair out of the range of the game, Cory settled in to watch the next round.

The remaining players abandoned their seats and another chair was removed. After Connie confirmed that there were now eight players and seven chairs, the music was started up again.

Emmy focused very carefully on the players this time, and began to gain confidence that this round would proceed more smoothly. As the music stopped, however, a confused and frustrated Marge was left without a chair, while Ivy seemed to be crouching on an imaginary chair.

Emmy looked quizzically at Ivy.

"My chair is invisible!" Ivy claimed.

Seizing the opportunity, Marge quickly crouched into a sitting position as well. "My chair is invisible too!" she countered.

Emmy looked to Connie once again for help. "I see only six chairs" Connie confirmed. "So there must be an invisible one somewhere."

"I can sort this out," Ivy's sister Vicky offered, and with a wave of her hand, the previously invisible chair appeared. But neither Ivy nor Marge was sitting on it! They both dashed to the chair, with Marge getting there first.

Ivy exited the game, shaking her head at her own mistake. "I made that chair invisible so that no one else could sit it in it," she explained to Emmy. "But then I forgot where it was!"

"Why do you all try to cheat at this game?" Emmy asked her. "It makes it very difficult to referee!"

"Trying to cheat is half the fun!" Ivy said with a wink. "It makes things much more exciting."

"I would say chaotic," Emmy countered. By this point, the remaining players were ready for the next round. There were now six chairs and seven players. As the music started again, Emmy had her eyes peeled for any mischief.

Everyone was still on their best behavior as the music stopped, until Rosalinda reached the last chair before Harvey and suddenly fell to the floor. As she struggled to right herself and understand what had happened, Harvey perched himself quite delicately on a tiny chair.

"You shrunk my chair!" Rosalinda accused Harvey.

"I shrunk *a chair*," Harvey insisted. "And now that chair happens to be mine!"

Emmy looked to Connie, who was shaking her head in disapproval.

"Disqualified!" Emmy pointed at Harvey.

"Oh fine," Harvey shrugged, and he scooted his tiny chair out of the way.

Now there were six players left. Cory, Ivy, and Harvey watched intently from the sidelines, ready to help spot any odd events.

When the music stopped this time, it wasn't hard at all to spot the oddity, as Leo was suddenly perched atop an elephant!

"My chair turned into an elephant!" Leo proclaimed in confusion. "Does that still count?"

All of the Funvillians laughed good-naturedly at his predicament. His chair had transformed into a lumbering, gray elephant shape with large ears and a narrow back. Leo was stretched out along its back, gripping its ears to stay on.

"It still counts," Emmy assured him, while looking pointedly at Constance, who had been left without a chair. Constance was in turn looking bemusedly at

Leo, impressed that he had managed to stay on while his chair had transformed.

"Worth a shot," Constance shrugged. She helped them to remove the elephant-shaped chair from the game space and then climbed up on it to watch the remainder of the game from a high vantage point.

Much to Emmy's surprise, the next two rounds passed peaceably, with Vicky and Marge each taking their eliminations with grace. But the calm didn't last long, as when the music stopped again, Leo and Ida were both (unsuccessfully) trying to climb onto an upside-down chair! Meanwhile, Rosalinda had taken the one remaining adjacent chair and was looking guiltily in another direction.

"No rotating the chairs!" Emmy decreed. "Rosalinda is disqualified!"

"Who, me?" Rosalinda feigned innocence.

"Yes, you!" Emmy insisted. "Put that chair back right-side-up, please."

"Ok, you got it," Rosalinda gave in, righting the chair and exiting the game. Only Leo and Ida were now left.

Emmy was relieved. She did not expect either Leo or Ida would try to cheat.

All of the disqualified participants were watching carefully as the final round unfolded. When the music stopped, Ida was slightly closer to the chair. She began to sit just as Leo arrived. To Emmy's shock and dismay, Leo grabbed the back of the chair and pulled it out from under Ida before she was fully settled, and she tumbled to the ground.

"Leo!" Emmy exclaimed. Leo was still gripping the chair with both hands, looking surprised at himself. Still on the floor, Ida was laughing.

"You're starting to get the hang of this!" Ida told Leo. "You cheat just like a Funvillian!"

"Not very subtle, though," Constance observed.

"To be fair," Ida countered, "the chair that turned into an elephant wasn't very subtle either!"

Constance shrugged.

Leo extended a hand and helped Ida up. "The winner!" he declared, raising Ida's hand in the air in celebration.

"I think it's my job to say that!" Emmy pointed out. "But yes, Ida wins!"

"That was fun!" Leo said to Emmy. "This party is great, and it has everything! Good games and ice cream, two of my favorite things!" But suddenly his excitement vanished. "My ice cream!" he exclaimed sadly. "I completely forgot about it. Surely it is melted by now."

Ida, who was still standing close by, happened to overhear him.

"You might want to check," she suggested, winking. "Sometimes things around here last longer than you might think."

# A DESSERT FOR ALL SEASONS

Leo rushed off to retrieve his abandoned ice cream.

"Why did you wink?" Emmy asked Ida. "Do you know something we don't? His ice cream has surely melted to a puddle by now."

"You'll see," Ida said mysteriously. In just a moment, Leo had bounded back to show Emmy his ice cream dish. The two remaining scoops had remained in perfect condition!

"It's . . . still . . . good," he managed to say between bites, "as if . . . I had never . . . left it!"

Emmy was puzzled.

"I wonder how this is possible!" Leo said more clearly, finishing the last of his impressively preserved dessert.

Just then another game was being organized.

"We're going to play hot potato next!" Cory announced. "Leo, you should come play!"

Putting down his spoon, Leo ran over to join the game.

After the game of musical chairs, Leo was not surprised to find that playing hot potato with Funvillians was not quite the same as playing it with his friends at home. Time was kept by an hourglass that Rosalinda rotated once at the last moment to give her an extra chance to unload the potato, while Cory of course copied the potato, leading to much confusion and amusement. Heather was also caught cheating when Leo couldn't manage to pass the suddenly much heavier potato!

Hot potato was followed by several lively rounds of Simon Says, and then Ida declared that it was time for cake.

"Cake!" Leo rejoiced, "I like cake almost as much as I like ice cream!"

Ida served Leo and Emmy each a generous slice of chocolate raspberry cake with chocolate icing. "Thanks," Emmy said, then took her first bite. "This is delicious!"

Leo nodded his head vigorously in agreement.

Still enthralled by the antics of the boisterous, playful Funvillians, Emmy distractedly dropped a bit of cake onto her dress.

"Oh no!" she exclaimed, trying to brush it off. "Now there is a chocolate and raspberry stain!" Her dress was a sunny yellow color, so the dark stain was quite

noticeable. "What can I do?" she asked Ida. "This is my favorite dress, but now it is ruined!"

"I can help!" Blake volunteered. He was next in line for cake, but stepped aside to offer his services.

Emmy was uncertain. "That didn't work too well for my notebook," she reminded him skeptically. "But I guess you can try."

Blake used his power on Emmy's dress, while Emmy closed her eyes. She was almost afraid to look. But when she opened her eyes again, the stain was gone, and the dress was restored to its pristine condition! She was very relieved.

"Thank you!" she smiled at Blake. "This time it's just what I needed!"

"I'm still sorry about your notebook," Blake said sincerely. "But I'm glad I could help fix your dress."

"It's really okay about the notebook," Emmy assured him. "I now realize it has all worked out for the best anyway. I've been writing in the extra space all about my exciting adventures here in Funville. Let me show you."

Emmy retrieved her notebook and handed it over for Blake to read. While the other party guests continued serving and eating their slices of cake, Blake was engrossed in Emmy's writing.

"This is really great!" he told her. "You are a very talented writer."

Emmy blushed. "Thanks," she said.

"You were wrong when you said you didn't have a power," he winked. "Being able to tell a great story is a truly wonderful gift!"

"Speaking of gifts," interjected Fay, seemingly appearing out of nowhere. "It's time to give Ida her presents!"

"Presents?" Ida asked, looking around. "I don't see any presents."

"What do you mean?" said Vicky, discreetly waving a hand behind her back and revealing the wall of stacked presents. "They're right here in front of you!"

"Indeed, how could I possibly have missed them?" Ida said in mock surprise. "I can't wait to find out what they are, but it will take me forever to open all of these boxes!"

"You don't have to open them to see what's inside," Ivy suggested. "Let me help you."

Ivy made the boxes invisible, revealing their contents, seemingly suspended in mid-air.

"These are great! Thank you everyone!" Ida said, admiring the wall of gifts.

Emmy and Leo found themselves awed once again by the magic around them.

Ida turned to Leo and asked, "So have you figured out how my power works yet?"

Leo paused to think carefully. He smiled as an idea occurred to him. "You can keep ice cream from melting!" he declared.

"That's one thing I can do," Ida admitted. "So I was the one who saved your sundae earlier."

"Thanks!" Leo said gratefully. "But what else can you do?"

"I told you already," she insisted. "My power is to keep things the same."

Leo was still puzzled.

"I'll give you another example," Ida offered. She reached for a balloon among the various party decorations. "What do you think will happen if I use my power on this balloon?" she asked Emmy and Leo, holding it by its string.

Emmy contemplated this for a moment, then exclaimed, "I bet it will never deflate!"

Ida nodded. She applied her power to the balloon and handed it to Leo. "Here," she said. "A small present for you. Take this home with you and you'll see!"

"Home!" Emmy remembered. "We still don't know how to *get* home!"

And for the first time since arriving in Funville, Leo shared his sister's concern. What if they couldn't figure it out?

# A THiNG TO REMEMBER

"Well, how did you get here in the first place?" Ida asked.

"We came down a slide on a playground at our school," Emmy recounted, "and at the bottom, we found ourselves on the slide at the playground here in Funville instead."

"Maybe if we go down the slide here again," Leo suggested, "we'll end up on the slide back home?"

Emmy shook her head. "We went down that slide already when we were looking for Doug, and nothing happened."

"Hmmm," Emmy concentrated on the problem. Suddenly, she smiled. "I have an idea," she announced. "Maybe if we want to return to where we've been, we should try to reverse our steps!"

"What do you mean?" Leo asked expectantly.

"Well, since we came down a slide to get here," Emmy prompted, "maybe we need to- "

"Go up the slide!" Leo finished. "Let's go try it!"

Emmy and Leo said goodbye to their new Funvillian friends, and returned to the playground. Reluctant to leave but anxious to get home, they started climbing up the slide. It was not very easy at first, as the clean, slippery slide was difficult to climb up. However, with much effort they managed to get almost to the top, at which point they magically discovered themselves at the bottom of the Thief.

Lucy, Jeremy, and the other children were gathered around, as if hardly any time had passed. Lucy's eyebrows were raised in dramatic fashion and Jeremy's face was seemingly frozen in its expression, with his jaw gaping wide.

"What happened to you guys? Where did you go?" All of the children began talking at once.

"I'm sooo sorry I dared you!" Lucy exclaimed. "I never thought in a million years that *you* could disappear!"

"It's ok," Emmy reassured her. "We had a grand adventure! We went to this whole other world and the people there had powers and . . . "

Suddenly Emmy realized the best way to explain.

"I wrote it all down!" She raised her notebook triumphantly. "I can read it to you."

"Yes!" Jeremy agreed, anxious to hear what had happened.

The children gathered round to hear Emmy read her story. Leo jumped in at several points to add details, like how good the ice cream tasted and how it felt to sit atop a chair as it became an elephant. "It was *this* big," he gestured grandly.

The other children were captivated by the story, and many wanted to try to go down the slide themselves. But then the bell rang, and they resolved to try again another day.

As the other children headed back to class, Emmy and Leo sat together for just a moment, thinking of their adventure and not wanting to accept that it was over. "I just had a scary thought," said Emmy. "What if you were high up in the sky in a hot-air balloon and Ida used her power on it. Then you'd never be able to get down!"

Leo laughed at the thought. "You would be stuck!"

This reminded him suddenly of the balloon. "Oh no!" he lamented. "The balloon that Ida gave me . . . I think I let go of it when we were climbing up the slide. Now I have nothing to remember Funville by."

At this point he put his hands into his pockets and sighed. After a moment though, he realized that there was something strange in his pocket that felt like a coin, but in an unfamiliar size. He took it out and looked at it carefully. It was a quarter, the quarter that was left over from his lunch money, yet something was different.

"Hey, this quarter is much smaller than usual!" he exclaimed, holding it up excitedly.

Emmy took it from him and examined it. "I bet it's exactly half the size of a regular quarter," she suggested slyly.

"Harvey!" exclaimed Leo. "It must have shrunk along with me. But then how come it didn't fully grow back like I did when Doug used his power?"

"I guess that will remain a mystery," said Emmy with a smile. "Just make sure not to lose it!"

"I won't," vowed Leo. "And maybe I will write my own story about Funville."

"And about whatever adventures we have next!" Emmy suggested.

Leo smiled at the thought.

# ADDENDUM

Dear reader,

Now that you have finished reading about Emmy and Leo's adventures in Funville, it is time to reveal to you the big secret behind this book - maybe you've already guessed - it was inspired by math! In fact, there are mathematical concepts hidden throughout the story. Are you ready to discover them?

We're going to use some mathematical terms now, but don't be worried if they are unfamiliar. We will define them for you as we go, and use examples from Funville and everyday life to illustrate them.

The main mathematical concept behind the story is that of a *function*. A function is just like a Funvillian's power that can be applied to certain objects (most of the functions that you encounter in school can be applied to various types of numbers). The set of objects that a function can be applied to is called its *domain.* For example, in Chapter 4 we discover that living things are not in the domains of Cory and Marge's powers.

Cory's left boot, however, is in the domain of his power, and he uses it as an *input* to his function, meaning that he applies his power to it. The result of this, which we call the *output*, is two left boots! This is because Cory's power makes a copy of whatever goes in, resulting in two identical objects as the result. Marge's power can then be used to take two identical objects and put them back into one.

You probably noticed that some Funvillians have siblings (like Cory and Marge) while others do not. Similarly, some functions can be *reversed*, while others cannot. When Harvey shrinks something down to half its size, Doug can reverse this effect by expanding it back to its original size. Doug's power is an *inverse* to Harvey's power: if Harvey applies his power to an object and then Doug applies his, the object ends up in the same state as it started. For Blake's power, however, there is no such inverse! Once information has been erased, it cannot be recovered. This is a *non-invertible function*: it has an effect on the object that can't be undone.

Constance's power is also non-invertible. In some cases, this can be very useful. Before she turned the playground springs into identical elephants, there was a red pig and a brown cow. Most Funvillians preferred the red pig, and as a result, they often argued over whose turn it was to ride it. Tired of the fighting, Constance turned them both into elephants. Soon, no one could remember which one had been the pig and which one had been the cow, and they no longer argued about turns.

Powers can also be *self-inverse.* For example, if Rosalinda uses her power of rotating things around half way twice, the object will be back in its initial state. Roberta, Robin, and Roy can also use their powers multiple times and get back to where they started. Such powers are called periodic. You've likely encountered *periodic* things in other settings, like the pattern of seasons repeating, or the sun rising and setting each day.

We also learn in Funville that powers can be applied to other powers, as Fay can change the powers of others. A function that changes other functions is called a *functional.* It may seem unnatural to apply a function to another function instead of a more basic input like a number, but why not? You could write a story about someone writing a story, for example, or take a picture of a picture, so why not make a function of functions?

If we take two Funvillians and they both use their powers on the same object, does it matter who goes first? The answer is, sometimes it does and sometimes it doesn't. For example, when you get dressed in the morning, it doesn't matter whether you put on your shirt first and then your pants or the other way around. On the other hand, you better put your socks on before your shoes or you'll end up with very dirty socks.

We say that the functions of putting on your shirt and putting on your pants *commute* whereas the functions of putting on your shoes and your socks *do not commute.* In the story we see an example of how knowing when functions commute can save you work when Emmy suggests to Cory and Marge that they blow up the balloons before copying them.

If you reflect on these examples (and we encourage you to come up with some of your own!), you will see that functions are a very important concept not only within mathematics, but they can also describe how we interact with our world. Whether we are putting on our socks and shoes, preparing food for a meal, or molding a ball of clay into a piece of pottery, we are seeking to move and transform objects, shaping them into the things we want them to be. Now that you understand functions, you are likely to see them everywhere! Recognizing and naming them will help you see the patterns in your daily experiences, and allow you to help devise creative solutions to problems you may face, just like Emmy and Leo.

Thanks for joining Emmy and Leo on their adventures in Funville!

A.O. Fradkin
A.B. Bishop

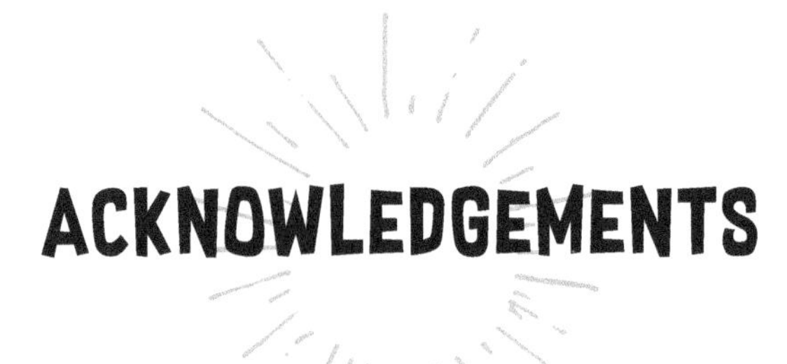

# ACKNOWLEDGEMENTS

We would like to thank our families and friends for their support throughout this project, as well as the early readers of the book and everyone who contributed to our Kickstarter campaign. We are grateful to Maria Droujkova and Natural Math for guiding us through the publishing process and helping us turn this book from a dream into a reality.

Anonymous
Abe Bassan
Abhishek Khurana
Ada and Linnaea Nesse
Adam
Adam and Ashley Bryant
Adamisho
Adamson family
Addy's Daddy
Adeline-Rose Williams
Adhya Mishra, Udayachal Primary School, Mumbai
Adina and Simona
Aditya Dhara, Gurgling Brook Foundation
Adrian Goh
Adriana Lopez-Alt
Agraj Jain
Ahilan
Ainsley family
Ainsworth family
Albert Chae
Aleksandra
Alena Kosik
Alexander & Galina F.
Alexander King
Alexandr Rozenfeld
Alexandra Hilderbrand
Alexandros Barbas
Alexey & Tania
Alexey and Uliana Makarov
Alexis and Anberlin Martinez
Alfie Exley
Alfreda Poteat
Ali Altug
Alison Davies

Alla Babayeva
Alla Proshina
Allison & Lizzie
Amit Sahai
Andrea Fant
Andrew Binder
Andrew Ng
Andrey and Irina Lvovsky
Andrey Fradkin
Andy R.
Angela DeHart
Anna Bekker and Maria
Anna Scholl
Anna Shershevsky
Anna Vinogradov
Annabelle Jane Woodruff
Anne Chen
Anon E. Moose
Anton Akio Sweetbaum
Ari Libova
Aria McAlister
arnoubea
Ash Fonarev
Asher and Isaac Rubinson
Asher Ling
Ashley Lesser
Atticus 顧樂毅 & Ariadne 顧樂盈
Audrey Whitworth
August & Theo Pompon
Averboukh family
Awesomesauce
Ayva Carrillo
Barbara Shaurette
Bear and Ayla Rosen
Ben Best
Betty d'Herve
Bill Blaskopf
Bill Farrow
Bluebaery
bogi_r79
Boris and Elina Ovetsky
Brett & Stephanie Vance
Brynn Saylor
Budsies LLC
Caelan, Ayalas, and the Pechatnikovs
Carbajal family
Carl
Carlos Santillan
Carly Espinas
Carol Cross, Heroic University
Carrie Aisen
Carrie House
Cassie Theobald
Catriona
Catriona Kaminski
CC7online
Charles Chen
Charlotte Ogilvie
Cheryl Ponchin
Christian A. Coker
Christopher Danielson
Christopher Frangione
Cinnia Hurley
Clément Canonne
Colin Murphy
Cooper & Danica Allen
Corinna Ulcigrai
Cory Doctorow
Craig Gentry, Michelle Montgomery, and Imogen Gentry
Craig Shelden
D'Arcy Brewer
Dan Hinderliter
Dana Bauer
Daniel Aisen
Daniel and David Jorza
Daniel Bauer
Daniel Wichs
Daniele
Danielle Nagelberg
Darlene Ing
Dasha Dougans
David & Eleanor
David Armstrong
David Petrie Moulton
David Pringle
David Ryvkin
Demi McCarthy
demona fett
Denis Savenkov
Denise Gaskins
Diego Morales
Dinah Slepovitch
DJH
DKCDA
Dmitriy & Elizabeth Vinogradov
Dmitriy Fradkin
Dmitry
Dmitry Bryazgin
Dodo, Uncle Brandon, Eden, and Aviva
Dom Groenveld
Dominic Choo
Dominique Charlebois
Doug Rindfleisch
Dr. Ken Newbury
Drew Couch
Edan Coben
Edith & Esther
Elaine Wah
Elana Reiser
Elena Fuchs
Elena Koldertsova
Eli and Ryan Weinberg
Eli Raymond
Elias and Jack Germany
Eliza and Tate Olsen
Ella Bokankowitz
Ellen Chang
Ellen Power
Elliott Feldman
Emiko and Nico
Emilia and Evan

Erwin Ari Martin
Esther and Franka
Ethan Bowden
Eugene B.
Eva Fridman
Evaleen & Ezeriah Perez-Lourido
Fatma Gurel Kazanci
Fauve Trudeau, Future Polymath
Fawkner Primary School Australia
@fearlessmath
Filip Skakun
Fox T.
Gabriel Birke
Gabriela Rodriguez Beron
Galeet Cohen
Geir Helleloid
Generation Mindful
Ghada Almashaqbeh
Gillian Grisenti
Gilman family
Giovanni Rosso
Girls' Angle
Golden Key Russian School
Grace Chou
Gracie Chae
Graham Bailey McMunn - Uncle Robert
H. Velasco
Hale family
Hanna Haydar
Hannah Tatro
Hannah Yoshiko Boas
Hannelore DK
Harper and Brynn Aion
Heidi & Thomas
Henry & Eddison Charno
Henry and Viola Yamane
Henry Cohn
Herr Sven
Hoste
Ida Knudsen
Iliya Koptsiovsky
Ilya Kaplun
Inga Zarecka
Irene Zubarev-Foxworth
Iryna Kadol
Isabella McLean
Iwona Hall
Jacob Tsimerman
James Gordon
James Tanton
Jan Kats
Jan T.
Janelle Campos
Jason Martindale
Jennifer Austin
Jenny Ratnovsky
Jeremy Secaur
Jeremy Smith
Jessica Jaskot
Jessie Oehrlein
Jo Oehrlein
Joe Gallian
Joe Kilian
Johann
Johanna Mareen
John Golden
John McAulay
Johnathan
Johnty Wang
Jojo
JOJO HOWELL
Jon Musky
Jon Wenger
Jonathan Pallie
Jordan Ellenberg
Joseph & Rex Ross
Joseph Meyer
Josh Giesbrecht
Joyce
Julia Bayuk
Julia Fuma
June R. Turner
Justin Simonson
K. Hughes
Kailee Auston
Kalashnikov family
Kara Dyer and Sara Argue
Karen Rothschild
Karen Toll
Karina Butterworth
Kasper Pater
Kathleen
Kathy Amen and family
Katie Hung
Katya Avdeeva
Keira Driansky
Kelly Vernon
Kelsey Williams
Ken Fan
Kenneth J. Carpenter
Kenny Shen
Kian Amiri
Kimberly Ichinose
Klamath KID Center
KO
Kulish-Maga family
L. Yeo
Lara Herrmann
Larry Wilson
Laura F. Cohen
Laura Green
Leo Sunday
Leonid Reyzin
Letfik
Liam Patterson
Libby Bishop
Lina Hasnai
Lina Kogan
Lisa Giaime
Livshins
Liz Berk
Logan & Sage Roe
Lohmann family
Lorelei Calderon
Lucas Garcia
Lucy Ravitch

Lumin & Irie Crewe
Lusann Yang
Lynne and Miller
Madeleine Ng
Madho
Mahdi Cheraghchi
Maia McCormick
Main Line Classical Academy
Manasson family
Margaret Evans
Maria Bedneau
Maria Gorbunova
Maria Schmidt
Marie Bishop
Marina
Marina Kopylova
Maritza Martinez
Mark G.
Mark Gurevich
Mark Shpilman
Martin
Martin Griggs
Marvin Hoheisel
Mason Boy
Math for Love
Math Hound
Math lover
Matt DeVos
Matt GS
Matt Jensen
Matthew
Matthew D. Reames
Matthew James Andrew
Matthew LaCoste
Matthew Waldschmitt
Matthew Yee
Maya Clark
Maya Vayner
Mayur Chauhan
Meghan Rose Allen
Melanie & Alex Bowditch
Melanie and Phil Wood
Michael and Robert Fouse
Michael J. Chon & family
Michael Li
Michael Ritzler
Michael Segal, to Isaac Sedighi
Michael Tarwater
Michael Vinogradov
Michael, Katie, and Lily Easton
Michelle Shwarz
Midgey and Little Man
Mika Parekh
Mike Hamburg
Mike Howard
Mike Smoot
Mike Teed
Mimi Rosen
Misha
mogo
Molly Gildea
Morgan Lota
MRS Productions
Ms. Zhang
Myles Scharf
Nancy K. Klein
Nandini Parikh
Nandini Ranjitkumar
Naomi Lara, Tomer, and Noga
Narahari Padayachee
Narasimha Raju Nagubhai
Nash LeTard
Natalie Hausknecht
Nataliya O.
Nathan Stephan
Nessie Van Loan
Nicholas Bathum
Nick Zielinski
Nicole Fradkin
Nicole Reardon, Solway family, Anderson family, Pilch family, Joshua Jackson, and Mundy Turner
Niko Wally
Oleksandr Prykhodko
Olga
Olga Arkhipova
Olga Avdyeyeva
Olga Finikova
Oliver Pratt
Orange Pi
ottoguy
P. Reddy
Paddock Girls
Paige Randall
Pasha
Pat Bishop
Patricia Bergh
Paul H. Curry
Paul Mostinski
Perry Waters
Peters Design Company
Phil Holland
Phil Whineray
Phoenix Garcia
PlayDiscoverLearn247.com
Polina Popova
Prerak Sanghvi
Prime Factor Math Circle
Project Rosie
Puneet Singhvi
Qasim Javed
Rachel Funk
Rafael Adrian Reyes-Matos
Rafael Panner
Rajiv Iyer
Ramamani Ramaraj
Ray Yang
Rebecca & Thomas Berg
Redditor Hopeless at Math
Regina Khersonsky

Reilly Moran
Remy
Rhiannon Jones White
Richard Berg
Rishi, Madhu, & Sagar
Rodi & Rachel Steinig
Roman Baranovic
Roman Gaisidis
Ron Fertig
Rosemarie Gibson
Russell Kuhns
Ryan Peterson
Safanov family
Salil Vadhan
Sam Lewin
Samuel and Kristi Roods
Sarah Irene Barry
Sarah Meiklejohn
Savio family
Saya Nirody
SchoolPlus Princeton
Serge-Eric Tremblay
Seth Bergenholtz
sglvin
Shalom Craimer
Sheel and Rujuta
Sheetal Walke
Shrenik Shah
Simo Muinonen
Simon Gregg
Simon Marmorek
SIMONE ETHERIDGE
Sivakumar Ganesan
Sizzy Bobby
Sloan
Som Liem
Sophia Lee
Sophie and Emma Jasik
Stan
Stephen Hartke
Steve and Carla Kaiser
Steve Lord
Steve Myers
Stifter family
Su Yi Oezata
Susie Marvin
Svet Ivantchev
Sylwia Maciulewicz
Taejas Subramanian
Tal Malkin
Tamara Rubakha
Tatiana Ter-Saakov
Taycher family
Team F
Team Tsou
Tessa & Oliver
Tessa, Douglas, Wendy, & Andrew
thatraja
The Arnsdorf family
The Baid family
The Banda-Berry family
The Barill family
The Bartz family
The Biryukov family
the Blackstons
The Bourgoin family
The Brisson family
The Brodsky family
The Brooks family
The Bueckert family
The Burdess family
The Cain family
The Candy family
The Chizhik family
The Cortez family
The Cudmore family
The Darke family
The DeSipio family
The Eberharts
The Ebersold family
The Ebersole family
The Elizondo family
The Farris family
The Fedorov family
The Fishman family
The Fords
The Frey-Romano family
The Grandparents Fridmans
The Greenfield family
The Gregory-Brown family
The Grinshpun family
The Hang family
The Hansen family
The Haspel family
The Hayashi family
The Hughes family
The Huston family
The Jewsbury family
The Kacheria family
The Katsuyama family
The Khankin family
The Khourys family
The Kolodziej family
The Krafcik family
The Kutin family
The Kydd family
The Lam family
The Lernout family
The Lyman family
The Madel family
The Manser family
The Marmur family
The Martinez family
The Maucher family
The McCormack family
The McGraw family
The McKeown family
The Menayan-Dence family
The Mendoza family
The Merupulas
The Nawrocki family
The Owladi family
The Pachowka family
The Parno family
The Pascual family
The Patel family
The Peoples family
The Perry family

The Pike Boys
The Pologruto family
The Popov family
The Protsko family
The Reyzin family
The Richter family
The Roussanov family
The Ryvkin family
The Shaffer family
The Shilliam family
The Shpilman family
The Shtutman family
The Simon family
The Skvortsov family
The Sorokin family
The Stratton family
The Swamy family
The Sze family
The Tahon family
The Tomlinson Trent family
The Walsh family
The Wardlaw-Bailey family
The Weaver family
The Williams family, C, L, R, A, & M
The Wood family
The Zhen family
Theo Burbridge-Kelly
Theoretical Computer Science
Thomas Rondeau
Timothy Chow
Timothy George
Tony Jebara
Tornatore family
Trevor Armstrong
Troy Wahl
Tyler Oliver
Uncle Aaron and Aunt Kate
Uncle Robert
Ursulet Flamand
Valentine Githler
Vanessa McWhorter-Cook
Victoria & Michael Drob
Vladimir Savikovsky
Wendy Christensen
Wesley Elias Saunders
Winston Luo
Xiang Zhang
Yannis Markou
Yao Yue
Yasha Berchenko-Kogan
Yelena McManaman
Yelena Samisheva
Yelisav family
Yerdua Arze
Yulia Shpilman
Yury Buslovich
Yury Kats
Yuyan Zimmerman
Yvonne Yeh
Zahabia Shahpurwala
Zoey & Zane Chong
Zoey Ramirez
Þórður Eiríksson

# ABOUT THE AUTHORS

**Sasha (A.O.) Fradkin** has loved math from an early age, and seeks to share that love of math with others. After receiving her PhD in mathematics from Princeton University, she worked for several years as a professional mathematician and taught enrichment math at the Golden Key Russian School to children ages 4-10. Currently, Sasha is the Head of Math at the Main Line Classical Academy, an elementary school in Bryn Mawr, PA. She develops their math curriculum and teaches children in grades K-5. She writes a blog, Musings of a Mathematical Mom, about her teaching as well as various math adventures with her two daughters, and enjoys pondering exciting and engaging ways to present the beauty of mathematics to young children.

**Allison (A.B.) Bishop** grew up with a passion for writing, and initially disliked math because it was presented as formulaic. She belatedly discovered the creative side of mathematics and science, and now sees it as a vital component of the curiosity that drives her life. She is currently a professor of computer science at Columbia University as well as a quantitative researcher at the Investors Exchange. She remains an avid fiction enthusiast and writer, and is always seeking new ways to expose young minds to creative mathematical thinking and fuel their scientific curiosity.

# ALSO AVAILABLE FROM NATURAL MATH

*Math Renaissance* is for teachers and parents of children ages six and up. The authors share their insights on how math experience might be improved at home, school, and math circle. It is based upon Rodi Steinig's experiences teaching math and leading math circles, and Rachel Steinig's experiences as a school student and homeschooler.

*Avoid Hard Work* gives a playful view on ten powerful problem-solving techniques. These techniques were first published by the Mathematical Association of America to help high school students with advanced math courses. We adapted the ten techniques and the sample problems for much younger children. The book is for parents, teachers, math circle leaders, and others who work with children ages three to ten.

*Camp Logic* is a book for teachers, parents, math circle leaders, and anyone who nurtures the intellectual development of children ages eight and up. You don't need any mathematical background at all to use these activities – all you need is a willingness to dig in and work toward solving problems, even when no obvious path to a solution presents itself. The games and activities in this book give students an informal, playful introduction to the very nature of mathematics and its underlying structure.

Available from NaturalMath.com and online book stores.
Published by Delta Stream Media, an imprint of Natural Math.
*Make math your own, to make your own math!*

CPSIA information can be obtained
at www.ICGtesting.com
Printed in the USA
LVHW07s0108220218
567515LV00019B/73/P